Fernando Coelho

Fighting Forest Fires to save water resources

Fernando Coelho

Fighting Forest Fires to save water resources

Determining the efficiency of chemical flame retardants in fighting forest fires

Imprint

Any brand names and product names mentioned in this book are subject to trademark, brand or patent protection and are trademarks or registered trademarks of their respective holders. The use of brand names, product names, common names, trade names, product descriptions etc. even without a particular marking in this work is in no way to be construed to mean that such names may be regarded as unrestricted in respect of trademark and brand protection legislation and could thus be used by anyone.

Cover image: www.ingimage.com

This book is a translation from the original published under ISBN 978-613-9-63518-4.

Publisher:
Sciencia Scripts
is a trademark of
Dodo Books Indian Ocean Ltd. and OmniScriptum S.R.L publishing group

120 High Road, East Finchley, London, N2 9ED, United Kingdom
Str. Armeneasca 28/1, office 1, Chisinau MD-2012, Republic of Moldova, Europe
Printed at: see last page
ISBN: 978-620-7-66735-2

SUMMARY

Forest fires today represent one of the main environmental problems facing the world, consuming thousands of hectares of forest every year and directly influencing global warming due to the carbon dioxide (CO_2) released in huge quantities by these fires, as well as being one of the biggest culprits in the destruction of natural animal habitats on the planet and, consequently, one of the biggest causes of species extinction. As a result, new technologies and procedures have been researched over time to control forest fires and combat them more efficiently, with less wear and tear on firefighters and less water consumption, given the increasing scarcity of this resource. As it becomes more difficult to obtain water and the intensity of the fire increases, the need arises to use chemical products that improve the efficiency of water in extinguishing the fire or that can replace it. The aim of this work was to optimise the Standard Operating Procedure (SOP) for fighting fire in vegetation using chemical flame retardants, with a view to saving water resources. The long-lasting flame retardant PHOS-CHEK G75R inhibited the germination of black bean seeds (Phaseolus vulgaris) at concentrations of 13.4 per cent, 6.7 per cent, 3.35 per cent and 1.675 per cent, so its use in drawing up a SOP for fighting forest fires was discarded. On the other hand, the short-lived retardant COUTOFLEX FIREKILL at a concentration of 1.125% showed germinations of black bean seeds (Phaseolus vulgaris) very close to those of the test samples. This compound performed satisfactorily in firefighting, being able to reduce the flame length by 20%, the speed of fire propagation by 42% and promoting a 40% reduction in flame intensity, even when used at a concentration of 1.125%. Based on these results, a SOP was drawn up with the aim of reducing the flame length below 1.5 metres, enabling the direct use of hand tools and thus reducing the consumption of water used to fight the fire.

Keywords: 1.Firefighting; 2.Forest Fire; 3.Water Resources; 4.Chemical Flame Retardants.

Thank you

I would like to thank my parents who have always supported me in my decisions.

I would also like to thank my friends: Erlon Paulo, Renato (Renatinho), Leandro (Lelê), Marcílio and Celebrim.

Finally, I would like to thank my supervisors Prof Dr Alexander Machado Cardoso and Prof Dr Gilberto Jorge da Cruz Araujo for their relevant guidance.

SUMMARY

CHAPTER 1   **4**

CHAPTER 2   **7**

CHAPTER 3   **28**

CHAPTER 4   **29**

CHAPTER 5   **36**

CHAPTER 6   **51**

CHAPTER 7   **53**

CHAPTER 1

INTRODUCTION

Throughout human evolution, fire has always been admired, feared and respected as a mystical element that modifies the environment and is capable of promoting the advancement and survival of the human species. Despite all the positive contributions that fire has made, it also has an extremely damaging aspect: fires (BOTELHO, 1996).

It should also be noted that fire plays a fundamental role in maintaining some ecosystems, since many trees and plants in the cerrado are already adapted to fire, some even needing it for some reproductive processes; however, its uncontrolled occurrence can represent a source of permanent disturbance, leading to irreparable material and environmental losses and damage (NUNES, 2005).

Forest fires are one of the world's most serious disasters. They consume thousands of hectares, causing major effects from both an environmental and economic point of view, as well as having a direct influence on global warming due to the release of large quantities of carbon dioxide (CO_2), one of the gases responsible for the greenhouse effect (BATISTA, 2004).

Forest fires also cause damage to fauna, either by killing them directly through the flames or indirectly through the changes caused to their ecosystem. Fires in vegetation are one of the main causes of the destruction of animals' natural habitats worldwide and, consequently, one of the main causes of species extinction (RODRIGUES, 2008).

In view of the annual losses caused by fire in vegetation, with the acceleration of human occupation in previously uninhabited areas and the increase in environmental and economic damage caused by forest fires, the need arose for new strategies, instruments and procedures to be sought in other countries and adapted in order to increase the efficiency of combating these incidents in Brazil (RIBEIRO, 2011).

In addition, the current water crisis that is plaguing almost the entire country is forcing us to look for new ways to combat it, taking into account the saving of water resources. The demand for water grows every year and the available reserves are exhaustible, so society needs to be alerted to the importance of saving water (use and reuse), with the establishment of public policies that guarantee the democratic, sustainable and integrated management of water resources (CALADO, 2015).

Faced with this problem, it is necessary to search for new technologies and/or new methods and procedures for fighting forest fires, with the aim of controlling them more quickly, more efficiently and with less wear and tear on the firefighter, taking into account the amount of water resources that will be consumed during the extinguishing of these fires. In Brazil, most forest fires are caused by direct human action, accounting for approximately 97 per cent of occurrences (SOARES, BATISTA and SANTOS, 2005).

There is no doubt that prevention is the best way to reduce the occurrence of forest fires. According to Vélez (2000), forest fire prevention is a set of measures aimed at reducing or cancelling the likelihood of a fire starting and limiting its effects if it does occur, i.e. preventing it from starting or hindering its spread.

Despite all the advances in fire monitoring systems aimed at preventing and spreading fires, in 2015 the Space Research Institute (INPE, 2016) recorded more than 235,000 outbreaks of fire in forest areas in the country, the second worst year since 1999, only lower than in 2010, when there were 249,000 fires.

This dissertation considered firefighting after a fire has started, looking for solutions to extinguish the flames so that they consume the least amount of vegetation possible. One of the ways to promote this efficient and water-saving firefighting is by implementing the use of chemical fire retardants in forest firefighting activities. The main characteristic of these products is their ability to reduce or even suppress flames by blocking the spread of fire. In Brazil, the use of chemical retardants is still little explored. These products offer great possibilities for fighting forest fires; however, there are still many uncertainties regarding their efficiency and operational capacity, due to a lack of knowledge about their behaviour in the environment, procedures and environmental toxicity (BATISTA, 2009).

The application of chemical retardants in firefighting actions aims to change the impossibility of fighting, creating another reality that allows the fire to be approached and extinguished (CUNHA, 2010), as well as reducing water consumption in firefighting.

This project sought, firstly, to verify the efficiency of these products and to study the possibility of expanding their use by the garrisons, optimising the Standard Operating Procedure (SOP) for fighting forest fires. This work presents a theoretical reference on forest fires, covering the use of short- and long-term chemical retardants, their action on vegetation, and some requirements for their use. Finally, it presents a practical test to verify their toxicity using bean germination as a quick and inexpensive test, as well as measuring their efficiency as fire retardants.

CHAPTER 2

LITERATURE REVIEW

2.1. FOREST FIRE

Castro et al. (2003) define forest fire as the combustion, uncontrolled in space and time, of combustible materials in forest areas. Nunes (2005) states that forest fire is the term used to define an uncontrolled fire that spreads freely and consumes the various types of combustible materials in a forest.

For the Food and Agriculture Organisation of the United Nations (UN, 2015), forest is:

> Area measuring more than 0.5 ha with trees greater than 5m in height and canopy cover of more than 10%, or trees capable of achieving these parameters in situ. This does not include land that is predominantly under agricultural or urban use.

According to the definitions above, we couldn't treat any fire in vegetation as a forest fire, so for the purposes of this study, we'll use a broader definition such as that of Soares and Santos (2002), who understand that a forest fire is the free or uncontrolled spread of fire in forests and other forms of vegetation. In this way, we will not only address forest fires according to this definition, but also other forms of fire in vegetation.

Forest fires are a problem that plagues not only Brazil, but also several countries around the world. According to Stangerlin et al. (2007), entire forests have been burnt down in different countries as a result of human carelessness or by nature itself, such as lightning strikes, causing irreparable damage to ecosystems and great harm to local populations and species.

In the winter and spring months, due to low rainfall, there are constant concerns about fires. During this period, fires are responsible for huge losses in the agricultural and forestry sectors and for disrupting natural ecosystems (PEREIRA, 2009). In Brazil, the situation worsens every year due to climate change, which makes periods of drought increasingly intense, increasing the extent of the burnt area and damaging the environment and society (SOARES and BATISTA, 2006).

2.1.2. Causes of forest fires

In terms of the nature of the cause, forest fires are classified in terms of their chemical, physical and biological nature. Forest fires of a chemical nature are those that originate through a chemical reaction. Forest fires of a physical nature occur when the fire is caused by some physical phenomenon or effect. Finally, forest fires of a biological nature occur when the fire is caused or stimulated by gases from archaea, bacteria, fermentations, among others (CBMERJ, 2014).

In terms of the nature of the causative agent, forest fires are classified as human-caused and natural-caused. Forest fires caused by a human agent are those whose origin is related to the actions of human beings, whether caused accidentally or not, and are currently the main cause of forest fires. Forest fires caused by a natural agent, on the other hand, are those caused by some natural phenomenon, without the influence of human will or error (CBMERJ, 2014).

2.1.3. Fire components

Fire is a phenomenon that produces heat in a combustible body in the presence of air. Once a fire has started, the heat generated by combustion will provide the energy needed to continue the process. Three basic elements are essential to start a fire: fuel, oxidiser and heat. Without one of these three elements, there is no fire (SILVA, 1998). According to Motta (2008), fire moves three basic elements:

- Combustible - anything that can combust (wood, paper, cloth, tow, paint, some metals, etc.);
- Combustant - is any element that, by chemically associating with the fuel, is capable of causing it to combust (oxygen is the main combustant);

- Heat (ignition temperature) - is the heat needed to start and continue the burning process, i.e. the temperature above which a fuel can burn.

As scientific advances progressed, it was realised that in addition to the three factors mentioned above, combustion required the presence of a fourth element: the chain reaction. With the chemical reaction of combustion, free radicals are formed that contain a high amount of energy and, in this way, these elements react with other molecules to form more free radicals and so on, in order to expand combustion (CBMERJ, 2014).

Nowadays, in addition to the fire triangle, we also have the fire tetrahedron theory which, as well as including fuel, comburent and heat, also considers the chain reaction to be a component of the fire, since in order for the fire to stay alight it is necessary for the flame to provide enough heat to continue burning the fuel (PIAUÍ, 2010).

2.1.4. Phases of combustion in forest fuels

Forest fuels are materials available in the environment that can ignite and burn. The drier the forest fuel, the more likely it is to burn faster. The greater the amount of combustible material being burnt, the greater the amount of heat released. The more heat that is released, the more the fire will spread and spread (SILVA, 1998).

The combustion of forestry material basically comprises three phases: pre-heating, the gaseous phase or distillation and the incandescence phase or carbonisation (MOTTA, 2008).

Pre-heating. This is the first stage in the combustion of forest fuels, in which the combustible material is dried, heated and partially distilled, but there are no flames yet. The heat eliminates the moisture in the material and continues to heat the fuel to ignition temperature. The volatile components move to the surface of the fuel and are expelled into the surrounding air. Initially these volatiles contain **large quantities of water vapour and**

some non-combustible **organic compounds** (MOTTA, 2008).

Gaseous phase or distillation. At this stage, the gases distilled from the wood catch fire and combust, producing flames and reaching temperatures of around 1250°C. At this stage of the combustion process, the gases are burning, but the fuel itself is not yet incandescent (BATISTA and SOARES, 2006).

Incandescence or carbonisation phase. In the third and final phase of combustion, the fuel itself is consumed and ash is formed. The heat is intense, but there are practically no flames or smoke. The amount of heat released in this phase depends on the type of fuel, but in general it can be said that 30 to 40 per cent of wood's combustion heat comes from its carbon content. The temperature is around 400 to 800°C (SCHUMACHER, BRUN and CALIL, 2005).

2.1.5. Combustion products

Combustion produces a series of products from the reaction of the fuel with the oxidiser. These products may or may not be visible. They include smoke, flame, heat and gases (CBMERJ, 2014).

Smoke. This is the result of incomplete combustion, in which small solid particles become visible, varying in colour according to the type of combustion. For example: white smoke indicates that combustion is more complete, with rapid fuel consumption and a good amount of oxidant; black smoke indicates that combustion is taking place at high temperatures, but with a lack of oxidant. When the smoke is yellow, purple or violet in colour, it indicates the presence of highly toxic gases (CBMERJ, 2014).

Flame. The visible part of combustion, the flame is part of the combustion zone in the gas phase, emitting light. It normally occurs in a combustion that takes place in an environment rich in oxidiser (PMESP, 2011).

Heat. A form of energy that raises the temperature generated by the transformation of other energy, through a physical or chemical process. It can be described as a condition of matter in motion, i.e. the movement or vibration of the molecules that make up matter. Molecules are constantly in motion. When a body is heated, the speed of the molecules increases and the heat (shown by the change in temperature) also increases. Heat is generated by the transformation of other forms of energy, namely: chemical energy, electrical energy, mechanical energy and nuclear energy (PMESP, 2006).

Gases. These result from the chemical modification of the fuel associated with the oxidiser. Combustion produces, among others, carbon monoxide (CO), carbon dioxide (CO_2) and hydrocyanic acid (HCN) (CBMERJ, 2014).

2.1.6. How fires spread

The spread of fire is due to various factors and can occur through: irradiation, conduction, convection and projection (CBMERJ, 2014).

Irradiation. The transmission of heat by waves of heat energy travelling through space. Heat waves propagate in all directions and the intensity with which bodies are affected increases or decreases as they are closer or further away from the heat source.
A warmer body emits waves of heat energy to a colder one until both are the same temperature (PMESP, 2006).

Driving. According to Oliveira **(2005): "conduction is the transfer of** heat through a solid body **from molecule to molecule". For** Schumacher, Brun and Calil (2005), conduction is the transfer of heat by direct contact with the heat source and, because wood is a poor conductor of heat, transfer by conduction is of little importance in forest fires.

Convection. This is the transfer of heat generally in an upward direction, i.e. by the displacement of heated liquid or gaseous masses. This transfer is due to the difference in density of fluids or the flow capacity of liquids, which occurs with the absorption or loss of heat (CBMERJ, 2014).

Projection. This is the movement or fall of objects (essentially solids) that are burning and can cause another outbreak of fire. In a forest fire, a log that rolls from the top of a burning hill to a lower, unburned location is an example of this form of propagation mechanism (CBMERJ, 2014).

Radiation plays a fundamental role in the spread of fire, as it is responsible for pre-heating the fuel, causing it to release flammable gases which then ignite. Convection generates a vertical and horizontal air current, which increases the size of the flame and promotes the entry of oxygen, favouring combustion. Conduction, on the other hand, plays a minor role in the spread of forest fires (MARTINS, 2010).

2.1.7. Types of forest fire

The most appropriate classification for defining types of forest fires is based on the degree of involvement of each stratum of forest fuel, from the mineral soil to the top of the trees, in the combustion process. In this case, fires are classified as underground fires, surface fires and crown fires (PIAUÍ, 2010).

Underground fires. These occur in the lower part of the ground. The combustible materials are formed from the decomposition of organic animal or plant matter. They do not normally have flames and are characterised by a high amount of heat, burn slowly and are difficult to locate, making them difficult to fight (CBMERJ, 2014).

Surface fires. These are characterised by the burning of dead undergrowth, such as herbaceous plants; the layer of leaves, branches, etc. that are mixed with the earth that covers the forest floor (serrapilheiras), as well as trunks and especially material that has undergone decomposition (humus). These fires do not cause significant damage to large trees; however, they are extremely damaging to undergrowth and young plants, especially for their regeneration. They occur in a variety of vegetation and are characteristic of Brazil and Latin

American countries (SILVA, 1998).

Canopy fires. These are fires that develop in the treetops, where the speed and intensity of the fire is greater and faster, due to the high wind circulation in these areas. These are fires that burn fuels above 1.80 metres in height, completely destroying the foliage of trees and generally killing them. With the exception of exceptional cases, such as lightning strikes, crown fires originate from surface fires (PIAUÍ, 2010). Their main characteristics are the high amount of heat generated and the difficulty and danger involved in fighting them (CBMERJ, 2014).

2.1.7. Parts of a forest fire

In order to improve and facilitate the communication process between firefighters and experts in the field, a terminology for the parts of a forest fire has been adopted and is widely used (Figure 1).

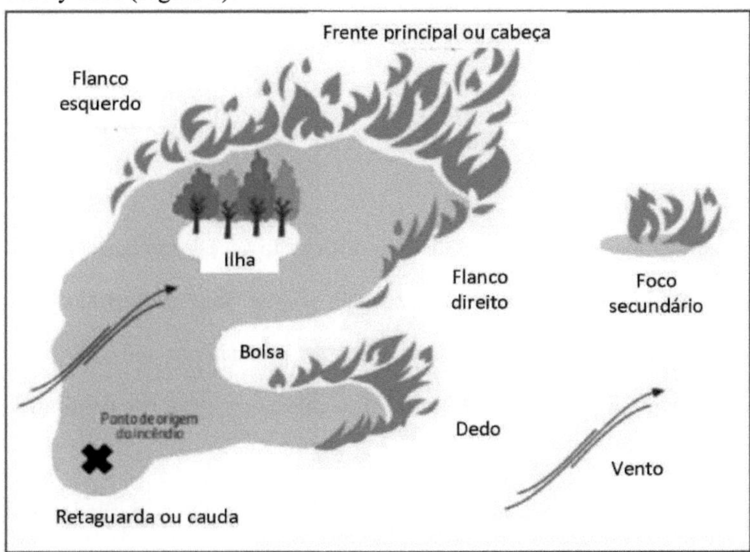

Figure 1. Parts of a forest fire.
Source: National Civil Protection Authority, 2009.

According to the Manual of the Military Fire Brigade of the State of Rio de Janeiro (2014), the parts of the fire are divided into:

-Main front   or head - the area where the fire spreads most intensely;

-Rear   or tail - the area opposite the front; where the fire is less intense, although it can also progress in that direction;

-Flank   - the side between the front and rear; the right flank is on the right side of the fire's direction of progress and the left flank is on the left side;

-Finger   - protrusion on a flank, corresponding to the place where the fire spreads most rapidly;

-Island - an   area within the perimeter of the fire that has not been affected by it, i.e. has not been burnt;

-Secondary fire - an external point, separated from the perimeter of the main fire, where a new fire ignites; and

-Pocket   - the area between the flank and the finger.

2.1.8   Materials and equipment used to fight forest fires

BATISTA and SOARES (2007 *apud* PARIZOTTO, 2006) define forest fire control materials and equipment as products or tools used to break down the combination of oxygen, heat and fuel.

Forest fires commonly occur in places that are difficult to access and cause a great deal of physical wear and tear on the combatants, which is why there are certain basic characteristics that all these materials must possess, such as: productivity and efficiency, versatility, portability, durability, simplicity, ease of maintenance and standardisation (CBMERJ, 2014).

Various types of materials and equipment are used by firefighters, brigades and volunteers all over the world to extinguish forest fires. Below are some of the materials and equipment most commonly used in Brazil to fight these fires.

Flexible backpack pump. The backpack or flexible backpack pump consists of a PVC tank with a capacity of approximately 19 litres of water and a quick coupling for the hose (CBMERJ, 2014).

Figure 2: Flexible backpack pump.
Source: Military Fire Brigade of the State of Rio de Janeiro, 2014.

Burner. The burners are metal containers with a capacity of 5 litres of fuel, which allow for dripping, with a fuel regulator.

It is used to reduce combustible material using fire or to initiate counter-fire techniques in firefighting operations (SANTA CATINA, 1994). It is used for techniques to reduce combustible material using fire or to initiate counter-fire techniques in combat operations (SANTA CATARINA, 1994).

Figure 3: Burner.

Source: Military Fire Brigade of the State of Rio de Janeiro, 2014.

Mcload. It is a versatile tool combining a hoe and a crawler in one piece, with high resistance. It is used for clearing vegetation, opening up small strips or firebreaks and digging small ditches (SANTA CATARINA, 1994).

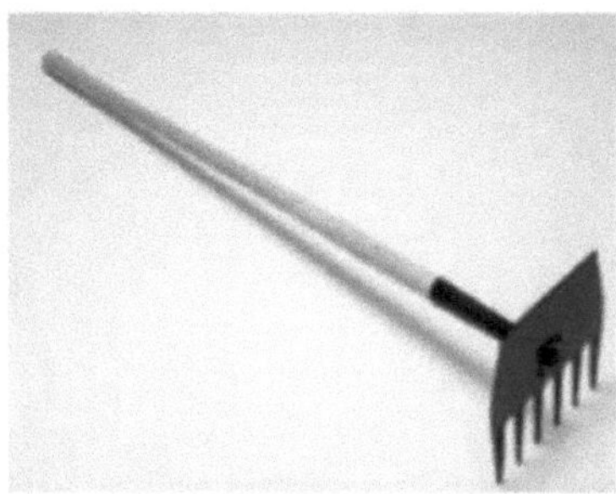

Figure 4: Mcload.

Source: Military Fire Brigade of the State of Rio de Janeiro, 2014.

Smother. It is used for direct firefighting by smothering and in the aftermath. It consists of a wooden handle, a rubber blade resistant to high temperatures and an iron support responsible for joining the handle and the rubber blade (CBMERJ, 2014).

Figure 5: Baffle.

Source: Military Fire Brigade of the State of Rio de Janeiro, 2014.

Chainsaws. Although they are not used exclusively to fight forest fires, chainsaws are also used in this type of event to saw down standing trees, prune branches or fallen trees, for example (SILVA, 1998).

Figure 6: Chainsaw.
Source: Solo, 2008, p. 162.

Scraping tools. According to SILVA (1998), these are tools that are more familiar to the general population on a daily basis, such as:

-Foice   : used to open access and cut branches;

-Hoe : used for weeding grasses, herbaceous plants and other types of undergrowth, as well as clearing trails;

-Crawler   : used to scrape the soil, removing plant fuels;

-Shovel   : used to mark out trees where the fire can pass, delimit the area for later firebreaks, cut branches and prune trees;

-Axe : used to cut down standing or fallen trees and to open lines and firebreaks;

-Shovels   : used for digging and throwing earth on the flames.

2.1.9   Methods of fighting forest fires

According to Motta (2008) there are three basic methods for fighting a forest fire: direct, indirect and parallel. In addition, for larger fires and/or those that are difficult to access, the aerial method can be used.

Direct method. According to Silva (1998), the direct method consists of fighting the fire directly, with dampers or by applying water, foam, retardant or earth.

Figure 7 - Direct combat method
Source: CBPMESP, 2006

Indirect method. A method in which strategies are used to contain and extinguish the fire without coming into contact with the flames. The construction of firebreaks (discontinuity in the vegetation) can be used to eliminate the fuel, or the rocky constitution of a given location can be used to trap the fire, or even the application of a cold line to moisten the vegetation, reducing the intensity of the flames, among various possibilities depending on the fire and the location (CBMERJ, 2014).

Figure 8 - Indirect combat method

Source: Instituto Alto Montanha da Serra Fina, 2013

Parallel method. This method is also known as the combined or mixed method. Rodrigues (2008) says that this method is used when the intensity of the fire or smoke is relatively high and combat personnel cannot approach. It consists of building a containment line just in front of the fire, so that when it advances along this line, its intensity decreases and it can be fought using the direct method. The containment line can be built using chemical retardants applied in front of the fire, or by mowing the vegetation and removing some of the combustible material. Both types of containment will reduce the intensity of the fire, allowing firefighters to approach.

Figure 9 -   Parallel combat method
Source: ICMBio, 2016.

Aerial method. According to the São Paulo Military Police Fire Brigade (2006), the aerial method can be used in areas or places that are difficult to access. This method is widely used in high-intensity canopy fires, using aeroplanes and helicopters that have been adapted or specially built to extinguish these fires.

Figura 10.   Air combat method

Source: Tomé M. and Borrego C., 2002.

2.1.10. Factors influencing forest fire behaviour

Predicting the behaviour and evolution of a given fire is based on an analysis of the factors that condition the fire: weather conditions, topography and fuels. To these variables must be added the factor of time, since the conditions of propagation change as the fire has a dynamic behaviour (VIEGAS, 2006).

The main characteristic of a forest fire is that it is unconfined and spreads freely. It is a fire that burns freely, responding to variations in the environment. Due to the chances of combinations of natural fuels, climate and topography, this fire can remain just a small smouldering spot or it can quickly develop into a large fire (SOARES, 1984).

Influence of topography. Within the topography factor we have some aspects that are very relevant and that contribute directly to the spread and planning of forest fire control, with exposure, slope and elevation being the most important (CBMERJ, 2014).

Exposure is an important factor because the sides facing the sun (or heat source) will

21

absorb more heat and have a higher temperature and lower relative humidity, making the material more vulnerable to fires (SANTOS, 2009). According to Castro et al. (2006), the greater or lesser inclination of a slope has a decisive influence on the spread of fires, due to the effect of convection columns that heat the vegetation above the fire. Thus, on a slope, the fire spreads much faster upwards than downwards. At higher altitudes, the air is thinner and temperatures are lower, so high altitude locations present less favourable conditions for the risk of forest fires (SAVIOLI, 1998).

Influence of meteorological conditions. The climatic conditions that most influence the spread of a forest fire are: Temperature, humidity and wind. Temperature is responsible for heating or cooling fuels which, at high temperatures, become drier and more vulnerable, especially when associated with low humidity. The humidity of the air and vegetation influences the spread of forest fires, because the higher the temperature, the greater the chance of water remaining in the atmosphere without condensing and, as a result, the humidity of the air remains higher. Similarly, at lower temperatures, the possibility of water vapour passing into the liquid state is greater, keeping less moisture in vapour form in the atmosphere. Atmospheric humidity also influences the humidity present in plants. During the day, the air removes moisture from vegetation, as it has a greater capacity to absorb water and a higher temperature than plants. The opposite happens at night, when plants absorb moisture from the cooler, more humid air. Finally, there are the winds which, as well as fuelling the fire by providing it with the necessary fuel, are responsible for directing the spread of the fire. Therefore, the wind direction must be constantly observed, because if it changes during the fight, a location that was once safe can become dangerous (CBMERJ, 2014).

Influence of fuels. Fuels in forest fires are plant biomass. The different species and configurations of the forest support the spread of fire in different ways (MARTINS, 2010). According to Piauí (2010), the spread of fire is usually faster and more intense in coniferous stands than in hardwoods (due to the presence of resin in conifers), faster in planted forests than in natural forests and even faster in pastures and dry fields. The fuel characteristics that most influence fire behaviour are: size and shape, quantity, humidity, continuity and compactness (BATISTA, 2007).

The behaviour of fire. Knowing how fire behaves is fundamental to guaranteeing safety conditions and organising efficient and effective firefighting. Basically, there are four parameters that govern fire behaviour: fire intensity, flame length, speed of spread and energy released during combustion (BATISTA, 2007). A forest fire is essentially a reflection of fire behaviour. The development of a fire, the effects it has on the soil and vegetation and the difficulty in controlling it depend on the behaviour of the fire (CRUZ and VIEGAS, 2001).

Fire intensity. Byram (1959) refers to the intensity of spread as the best indicator of fire behaviour. Alexander (2000) establishes intensity classes that are related to the height of the flame and the effectiveness of the firefighting means, and this relationship is a reference in the field of firefighting. Fire intensity was defined by Byram (1959) as the release of energy per unit time and per unit length of the fire front and its estimate can be obtained using the equation:

Equation 1: BYRAM equation for determining fire intensity.

$$I = H.\ w.\ r$$

Where:

I = Fire intensity in kcal.m^{-1} .s ;$^{-1}$

H = calorific value in kcal.kg^{-1} ;

w = Weight of combustible material in kg.m^{-2} ;

r = Speed of fire spread in m.s^{-1} .

Flame length. According to Batista (2007), flame length is the distance between the tip of the flame and the ground or the surface of the fuel that is burning, measured in the middle of its active zone.

According to Byram (1959), intensity is directly related to flame length (hc). The relationship between fire intensity and flame length is given by the following equation:

Equation 2: Relationship between fire intensity and flame length.

$$I = 63.05 \cdot hc^{2,17}$$

Where:

I = Fire intensity in $kcal.m^{-1}.s^{-1}$;

hc = flame length in m.

This is considered an alternative way of obtaining fire intensity.

Table 1, drawn up by Parizotto (2006), shows the relationship between the intensity of the fire and the length of the flames and the way in which the forest fire is fought.

Table 1. Relationship between fire intensity and flame length, and the way in which the fire was fought.

Length of flames (m)	Intensity fire ($kcal.m^{-1}.s$)$^{-1}$	Information on fire behaviour and fire-fighting methods.
< 1,2	< 80	Fires can usually be fought directly, at the head or on the flanks using hand tools. Small manual firebreaks are enough to hold the fire at bay
1,2 a 2,4	80 a 400	The fires are too intense to use the direct method, and manual firebreaks can't hold the fire at bay. Water pumping equipment and tractors with blades are needed to fight the fire.
2,4 a 3,3	400 a 800	Fires can present serious difficulties in being controlled and fought, as they cause the canopy to burn, and with it a lot of sparks.
> 3,3	> 800	These are extremely violent fires, with total burning of the forest and intense sparking, nothing can be done in front of the fire. You have to wait for a reduction in the intensity of the fire, usually caused by climate change.

Source: PARIZOTTO (2006, p. 41-42).

The Manual de un Primer Ataque a un Incendio Forestal (2002) relates the length of the flames to the recommended options for fighting a forest fire, according to Table 2.

Table 2. Relationship between flame length and recommended firefighting options.

Flame length (m)	Recommended combat options

<1,5	Direct firefighting with hand tools and, when the topography of the terrain allows it, firefighting vehicles on the head and flanks are sufficient for control.
1,5 a 2,5	The intensity of the fire makes it difficult to approach it directly with hand tools. When the topography of the terrain allows, firefighting vehicles and aerial means are suitable for fighting the blaze, in addition to the possibility of indirect firefighting.
2,5 a 3,5	Apply indirect firefighting strategies at the front of the fire, in addition to the use of aerial resources. The heat will not allow you to approach within 10 metres of the fire line.
> 3,5	Apply indirect fire-fighting strategies at the front of the fire, and in extreme cases, where there is a risk of death and/or damage to installations, the technique of counter-fire is a possibility, in addition to the use of aerial means.

Source: Adapted from Manual de un Primer Ataque a un Incendio Forestal, 2002.

Fire spread rate. Indicates the rate of advance of the fire in a given period of time over an area that is burning in a forest fire. It is an important parameter to evaluate because, by measuring the speed of fire spread, we can get an idea of the time we will have to prepare fire-fighting actions, such as building firebreaks (PIAUÍ, 2010).

Energy released in combustion. This is the total amount of energy released per unit area. During the period of combustion, it can be estimated through the intensity of the fire and the speed at which the fire spreads (BATISTA, 2007). This is a parameter that is widely used to assess the effects of fire on the soil, as the faster the fire spreads, the less heat is concentrated per unit area.

2.1.11. Chemical retardants used in forest firefighting

A chemical retardant is a chemical agent that can be used, alone or mixed with water, to reduce or eliminate the combustion of a given fuel. One of the biggest advantages of using these products is the increased efficiency of water use at a relatively low cost (SANTANNA, FIEDLER and MINETTE, 2007). When chemical retardants are applied directly to vegetation, there is a reduction in flame length and intensity. Many retardants also have a fire-inhibiting action, caused by the deposition of the product on the foliage of the vegetation, which can lead to fire suppression (CUNHA, 2010).

Considering the duration of the product's action, retardants can be classified into:

short-acting retardants and long-acting retardants (RIBEIRO et al., 2006).

Short-lived retardants are products that, when mixed with water, form a foam that increases its efficiency in extinguishing forest fires by up to five times (CUNHA, 2010). They contain a combination of surfactants that significantly reduce the surface tension of water, helping water to penetrate the combustible material. The foam produced has a certain adhesion to the vegetation and is very persistent (GUARANY, 2011).

One of these types of retardants is COUTOFLEX FIREKILL AFFF and ARC Foam Generating Liquid (LGE), a synthetic product that is low in toxicity, biodegradable and non-corrosive to metallic surfaces, according to the

The product is made up of water, butylcarbitol, alkyl glucoside, amphoteric fluorinated polymer and hexahydro-1,3,5-trios (2-hydroxyethyl)s-triazine. When applied, it forms a foam cover, which cools and extinguishes the fuel in order to minimise the evaporation of gases and prevent them from mixing with oxygen in the air. This foam has physical-chemical characteristics of resistance to high temperatures, when mixed with water, and spreads to form a cover over the fuel until the fire is completely extinguished, resisting the destructive effects of the heat radiated by the remaining fire, sealing the vapours with the action of fluorinated active surfactants, as well as being able to contain flammable gases and minimise the risk of re-ignition. As LGE contains 94% water in its liquid phase, it also acts as a heat exchange medium, helping with cooling, given that the product is used in concentrations of 9% (COUTOFLEX INDUSTRIA DE MANgueira, 2010).

LGE is normally used to fight class B fires (liquid fuels), such as: ethanol, petrol and diesel oil, for example. In Brazil, this type of flame retardant is rarely used to fight vegetation fires.

Long-lasting retardants are basically composed of ammonium polyphosphates, which are diluted in water for application. They can be applied directly to the combustible material to delay or even prevent combustion, and their action leaves residues of combustion-inhibiting agents on the combustible material even after all the water has evaporated (PARDO, 2007).

PHOS-CHEK G75R is a retardant classified as long-lasting and is a mixture of ammonium phosphate and ammonium sulphate. This product alters the flammability of combustible material. In the presence of the retardant, the release of flammable gases, which contribute to preheating and combustion, does not occur or is hindered. As the water evaporates, it absorbs heat, cooling the fuel and consequently making it difficult for the combustion reaction to continue because of the barrier formed (MONSANTO COMPANY). According to the manufacturer, the active ingredients used in the formulation are non-toxic to humans and the environment and easily biodegradable. It is recommended to use the product in concentrations of 13.4 %.

CHAPTER 3

OBJECTIVES

3.1 . General Objective

The aim of this dissertation is to evaluate the efficiency of chemical retardants in forest firefighting, verifying through tests the efficiency of these retardants and their effects on reducing the intensity of fire in vegetation, with a view to saving water resources in firefighting.

3.2 Specific objectives

- Check the toxicity of the chemical retardants PHOS-CHEK G75R and COUTOFLEX FIREKILL AFFF and ARC at the concentrations tested;

- To examine the efficiency of the chemical retardants PHOS-CHEK G75R and COUTOFLEX FIREKILL AFFF and ARC in reducing the speed of fire spread and in reducing the length and intensity of the flames;

- Find out which of the chemical retardants analysed is the most efficient and environmentally acceptable;

- Optimise the Standard Operating Procedure for fighting fire in vegetation by introducing the use of chemical retardants that improve the efficiency of the process and consequently result in the saving of water resources and the preservation of as much vegetation as possible.

CHAPTER 4

MATERIAL AND METHODS

4.1 Toxicity test for COUTOFLEX FIREKILL AFFF and ARC **flame retardants** and PHOS-CHEK G75R **to be carried out easily in forest firefighting units**

As this is a project that aims to find ways of extinguishing fires without damaging the environment, it would make no sense to use a flame retardant that is toxic to the environment. Therefore, aqueous solutions of the flame retardants PHOS-CHEK G75R, which is non-toxic to humans and the environment according to the manufacturer, and COUTOFLEX FIREKILL AFFF and ARC, which is low in toxicity according to the manufacturer, were used in different concentrations to check their effects on inhibiting the germination of black beans (Phaseolus vulgaris) used as a test organism. In the experiment, the beans were exposed to an acute toxicity test which, according to Duffus (1993), are experimental studies carried out with test organisms that determine whether an observed adverse effect occurs in a short space of time (usually up to 14 days) after administration of a single dose of the substance tested or after multiple doses administered within 24 hours.

The use of bean sprouts is based on studies that highlight the advantages of using these plants as test organisms for bioassays, such as economic viability, because they are low-cost tests; the ease of understanding how organisms react to toxic effects; and because they have a connection with bioassays that use animals as test organisms, making it possible to base the results of tests with plants on other living beings (FISKESJO, 1985).

Four aqueous solutions of different concentrations were prepared for each of the flame retardants tested. The concentrations used for the PHOS-CHEK G75R flame retardant were 13.4% (recommended by the manufacturer), 6.7%, 3.35% and 1.675%. The concentrations used for the COUTOFLEX FIREKILL AFFF and ARC flame retardants were 9% (recommended by the manufacturer), 4.5%, 2.25% and 1.125%.Table 3 shows the concentrations and quantity of product used in each test.

Table 3. Toxicity test of the chemical flame retardants Phos-Chek G75R and Coutoflex Firekill using black beans (Phaseolus vulgaris) as the test organism.

Test I			
Essay	**Experiment**	**Concentration**	**Qty of product/volume of H2O**
T	10 grains that received only water		
A	i0 grains in contact with Phos-Chek G75R	13,4%	26.8g/200ml
B	10 grains in contact with Coutoflex Firekill	9%	18ml/200ml
Essay II			
Sample	**Experiment**	**Concentration**	**Qty of product/volume of H2O**
Ti	10 grains that received only water	-	-
C	10 grains in contact with Phos-Chek G75R	13,4%	26.8g/200ml
D	10 grains in contact with Coutoflex Firekill	9%	18ml/200ml
E	10 grains in contact with Phos-Chek G75R	6,7%	13.4g/200ml
F	10 grains in contact with Coutoflex Firekill	4,5%	9ml/200ml
Essay III			
Sample	**Experiment**	**Concentration**	**Qty of product/volume of H2O**
T2	10 grains that received only water	-	-
G	10 grains in contact with Phos-Chek G75R	3,35%	6.7g/200ml
H	10 grains in contact with Coutoflex Firekill	2,25%	4.5ml/200ml
I	10 grains in contact with Phos-Chek G75R	1,675%	3.35g/200ml
J	10 grains in contact with Coutoflex Firekill	1,125%	2.25ml/200ml

In each trial, 10 beans were placed on a commercial cotton pad in disposable plastic containers containing 200 ml of water and the flame retardants at the predetermined concentrations. The containers were kept in an airy place exposed to sunlight for 14 days. The water absorbed and/or lost through transpiration and evaporation was replaced daily. These tests were carried out at the Seventh Military Fire Brigade in a rudimentary manner in order to demonstrate that this is a simple test that can be carried out in any forest firefighting unit.

4.2 Toxicity test of COUTOFLEX FIREKILL AFFF and ARC and PHOS-CHEK G75R flame retardants carried out in the laboratory

In order to make the study reliable, the above tests were repeated independently in the laboratory. In addition, germination tests were carried out on different bean varieties (white, carioca, black-eyed pea, butter bean, brown, mulatto, black, red) in the presence of the flame retardants COUTOFLEX FIREKILL AFFF and ARC and PHOS-CHEK G75R at the concentrations of 9% and 13.4% (indicated by the manufacturer), respectively.

4.3 Description of the experiment to evaluate the efficiency of the short-lived chemical flame retardant

The experiment was carried out at the Seventh **Military** Fire Brigade**, located at 22°31'48"S and 44°1T09"O, in the municipality of Barra** Mansa/RJ, inside the burning container (Figure 23) in order to minimise external effects, such as wind. On this date, the average ambient temperature was 28°C and the relative humidity was around 66%.

Figure 11. Container used to carry out the experiment to evaluate the efficiency of the Coutoflex Firekill AFFF and ARC chemical retardants.

The product analysed was the short-lived flame retardant COUTOFLEX FIREKILL AFFF and ARC. It was initially used at the concentration indicated by the manufacturer, 9%, with 450 ml of the product diluted in 5.0 litres of water. The mixture was then diluted to a concentration of 1.125% (the concentration obtained in the experiment to test the toxicity of the product), with 56.25 ml of the product diluted in 5.0 litres of water. The mixture was made directly in the backpack in order to be as close to the real thing as possible.

For each concentration tested, three burns were carried out on the burning platform, which is 4.0 metres long by 1.0 metre wide and divided as shown in Figure 12.

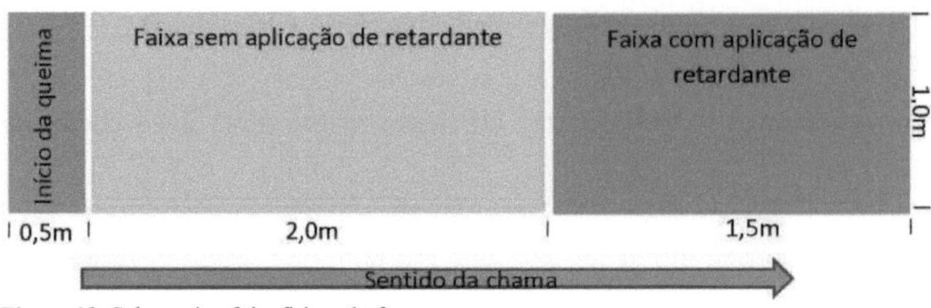

Figure 12. Schematic of the firing platform.

Markers were used to demarcate the strips without the application of the product, where burning occurs freely, and with the application of the product, where a reduction in the length and speed of flame propagation is expected due to the action of the retardant (figure 13).

Figure 13 - Use of markers to demarcate lanes.

To standardise the research, plots were set up on the burning platform with a concentration of 17.5 Kg.m^{-2} of combustible material with a blanket thickness of 25 cm. The fuel used was dried Brachiaria grass (Brachiaria spp), harvested on the same date and in the same place, in order to minimise fuel interference in the results obtained. In all plots, a mass of 70 kg of fuel was evenly distributed over the burning platform, as shown in Figure 14.

Figure 14 - Plots of fuel distributed over the firing platform.

For each burn carried out, 600 ml of the flame retardant mixture was applied to one side of the plot with water at the concentration tested, in the allotted space (1.5 x 1.0m), using the same nozzle flow rate for all applications, to ensure the most homogeneous application possible. The dosage of mixture was 0.4 litres per square metre of area in each burn and the application was made using a backpack. Two burnings were carried out for each concentration tested, totalling 06 burnings and measuring the following variables: propagation speed and flame length.

These variables were measured taking into account the intervals of the strip free of product application (2.0 metre strip) and the 1.5 metre strip with retardant application.

A lighter was used to start the blaze and after the fire had travelled the first 0.5m, measurements began to be taken.

The burns were carried out immediately after the application of the products, with the aim of simulating the action of firefighters or a fire brigade when fighting forest fires.

During the tests, a comparison was made between flame propagation speed in the area without the product and in the area where it was applied. The same analysis was carried out in relation to flame lengths, which is the most important parameter in this study, because according to the Manual for a Forest Fire Primer (2002), for flame lengths of up to 1.5 metres, direct firefighting can be carried out with hand tools, and according to Parizzoto (2006), for flame lengths of up to 1.2 metres, fires can be fought directly, at the head or on the flanks using hand tools or small manual firebreaks. Taking this information into account, if the application of retardants is sufficient to reduce the flame length to the point where only hand tools can be used, we can optimise the Standard Operating Procedure for fighting forest fires with the aim of saving water resources.

The speed at which the fire spread (m.s^{-1}) was calculated by determining the average time taken by the fire front to travel the pre-established distances during the burns. Flame length (m) was calculated using the average of the values collected during the burn. A strip determining the length of 1.5 m was placed next to the platform (figure 15), so that later, by analysing images of the experiment, it was possible to determine the length of the flames using a simple proportion.

Figure 15. Range used to measure flame length.
Source: by the author.

With the flame length data (hc), it was possible to obtain the reduction in flame intensity (I) using equation 2, which relates the two variables.

Equation 2: relationship between flame intensity and length

$$I = 63.05 \cdot hc^{2,17}$$

The determination of the reduction in flame intensity was calculated using the ratio between the fire intensity values obtained from the fires

carried out with the application of retardant with the respective results of this variable obtained in the burnings without the application of the product.

As seen above, the variables analysed were the speed at which the fire spread and the length of the flames. After carrying out the experiment and collecting the data, we moved on to the analysis phase.

CHAPTER 5

RESULTS AND DISCUSSION

5.1 Toxicity test for COUTOFLEX FIREKILL AFFF and ARC **flame retardants** and PHOS-CHEK G75R **to be carried out easily in forest firefighting units**

Figure 16 shows samples T (control), 1 and 2 on 21 October 2016, when the experiment began.

Figure 16. Photos showing the initial appearance of the black bean samples in the test using water only (T), a 13.4% aqueous solution of Phos Check G75 R (A) and a 9% aqueous solution of Coutoflex Firekill (B).

After 14 days of experimentation, at the end of trial I, it was found that only the black bean sample that received only water had germinated 06 grains, while the others had not germinated, as can be seen in figure 17.

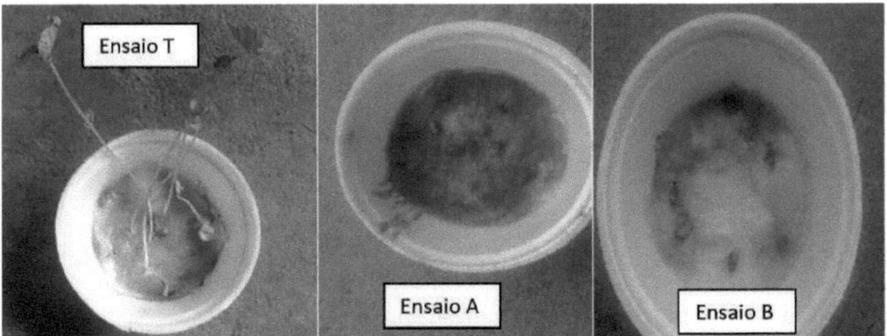

Figure 17. Photos showing the appearance of the black bean samples at the end of test I (after 14 days) using only water (T) a 13.4% aqueous solution of Phos Check G75 R (A) and a 9% aqueous solution of Coutoflex Firekill (B).

At the end of Trial II (14 days), which was carried out as described in Table 3, it was observed that 7 beans had germinated in the black bean samples that had only received water

(Figure 18). In the samples where aqueous solutions of Coutoflex Firekill were used at a concentration of 9% and 4.5%, there was slight germination of 01 and 04 beans, respectively (figure 19). In the samples where aqueous solutions were used with the flame retardant Phos Check G75R at a concentration of 13.4% and 6.7%, no germination was observed (Figure 20).

Figura 18. Photo showing the appearance of the black bean sample at the end of Trial II (14 days) which received only water (TI).

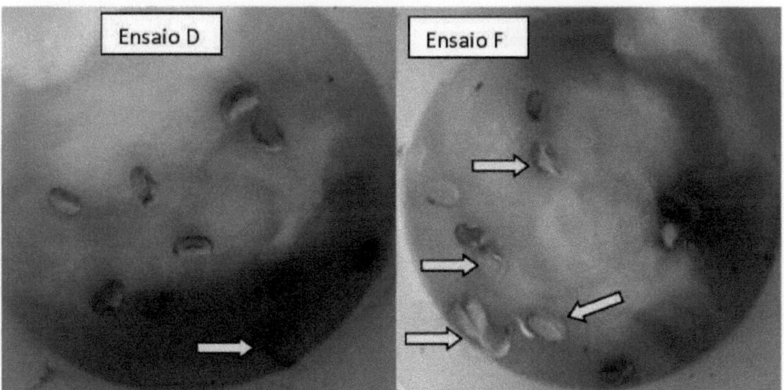

Figura 19. Photo showing the appearance of the black bean samples at the end of Trial II (14 days) that received the aqueous solution of Coutoflex Firekill at 9% (D) and 4.5% (F).

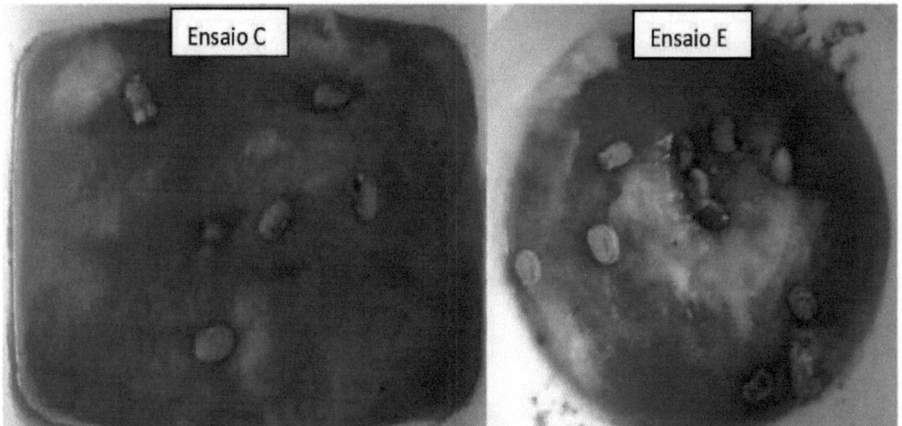

Figure 20. Photo showing the appearance of the black bean samples at the end of Trial II (14 days) that received the aqueous solution of Phos Check G75R at a concentration of 13.4% (C) and 6.7% (E).

At the end of Trial III, 14 days after the start, it was observed that 6 beans had germinated in the control sample (T2) of black beans that received only water (figure 21). In the sample where an aqueous solution of Coutoflex Firekill at a concentration of 2.25% was used, there was slight germination of 4 beans.

In the sample where an aqueous solution of Coutoflex Firekill with a concentration of 1.125% was used, 10 grains germinated (figure 22).

No germination was observed in the samples containing aqueous solutions of the flame retardant Phos Check G75R at concentrations of 3.35% and 1.675% (Figure 23).

Figura 21. Photo showing the appearance of the black bean sample at the end of Trial III (14 days) which received only water (T2).

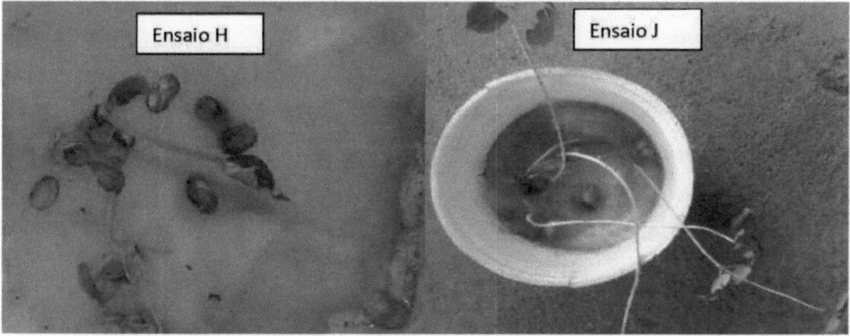

Figura 22. Photo showing the appearance of the black bean samples at the end of Trial III (14 days) that received the aqueous solution of Coutoflex Firekill at 2.25% (H) and 1.125% (J).

Figura 23. Photo showing the appearance of the black bean samples at the end of Trial III (14 days) that received the aqueous solution of Phos Check G75R at a concentration of 3.35% (G) and 1.675% (I).

Table 4 shows the number of seeds germinated in each sample in the tests carried out.

Table 4. Number of seeds germinated in each sample in tests I, II and III using water or aqueous solutions of chemical flame retardants.

Sample	Number of germinated seeds
T	6
A	0
B	0
T1	7
C	0
D	1
E	0
F	3
T2	6
G	0
H	4
I	0
J	5

n=10

Caption:

T: 10 grains that received only water

A: 10 grains in contact with an aqueous solution of Phos-Chek G75R at a concentration of 13.4%

B: 10 grains in contact with a 9% aqueous solution of Coutoflex Firekill

T1: 10 grains that received only water

C: 10 grains in contact with an aqueous solution of Phos-Chek G75R with a concentration of 13.4% D: 10 grains in contact with an aqueous solution of Coutoflex Firekill with a concentration of 9% E: 10 grains in contact with an aqueous solution of Phos-Chek G75R with a concentration of 6.7% F: 10 grains in contact with an aqueous solution of Coutoflex Firekill with a concentration of 4.5% T2: 10 grains that received only water

G: 10 grains in contact with an aqueous solution of Phos-Chek G75R with a concentration of 3.35% H: 10 grains in contact with an aqueous solution of Coutoflex Firekill with a concentration of 2.25% I: 10 grains in contact with an aqueous solution of Phos-Chek G75R with a concentration of 1.675% J: 10 grains in contact with an aqueous solution of Coutoflex Firekill with a concentration of 1.125%

Analysing Table 4, it can be seen that the concentration of the flame retardant COUTOFLEX FIREKILL AFFF and ARC has a direct link with the number of germinated seeds. The lower the concentration of the product, the higher the number of germinated seeds. The flame retardant PHOS-CHEK G75R, on the other hand, did not allow any kernels to germinate at any concentration tested. The graph below (figure 24) illustrates the relationship between the number of germinated seeds and the aqueous solution used.

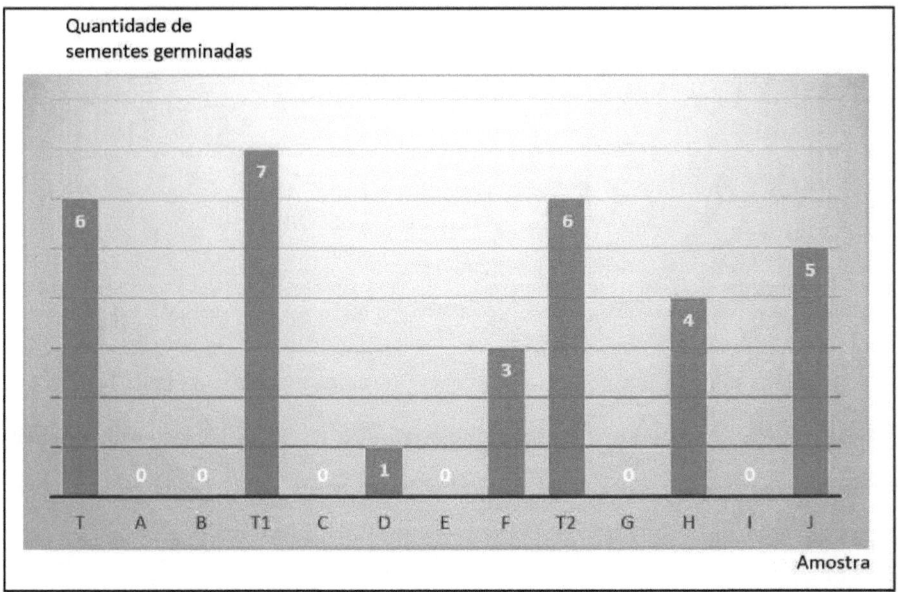

Figure 24. Graph showing the relationship between the percentage of Coutoflex Firekill AFFF and ARC flame retardant and Phos Check G75R used in the aqueous solution and the number of germinated seeds.

5.2 Laboratory toxicity test of COUTOFLEX FIREKILL AFFF and ARC and PHOS-CHEK G75R flame retardants

The results obtained in the laboratory tests were similar to those obtained in the simpler test carried out at the Seventh Military Fire Brigade. The flame retardant Phos Check G75R did not allow any black bean kernels to germinate at any of the concentrations tested and the flame retardant Coutoflex Firekill only showed satisfactory kernel germination results, coming very close to those obtained in the test samples, at a concentration of 1.125% (figure 25).

41

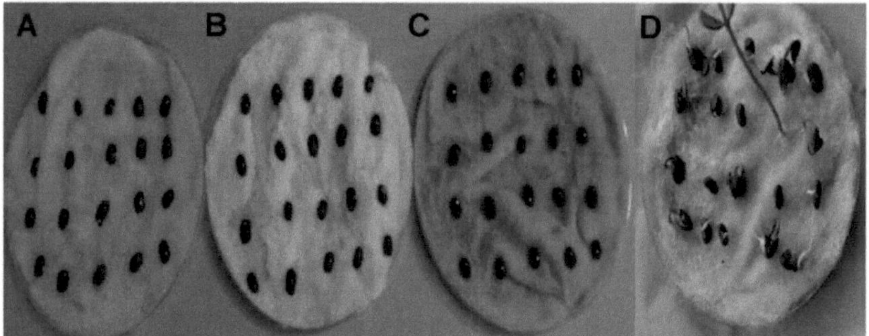

Figure 25. Germination test of bean seeds in the presence of the flame retardants COUTOFLEX FIREKILL (B) and PHOS-CHEK G75R (C) compared to the control (A). Germination only occurred in the control group (D).

Aqueous solutions of the flame retardants Coutoflex Firekill and Phos Check G75R at 9% and 13.4% concentration, respectively, were also tested in contact with different varieties of beans (white, carioca, black-eyed peas, butter beans, brown beans, mulatto beans and red beans), yielding the same germination results obtained when using black beans: no germination (figure 26).

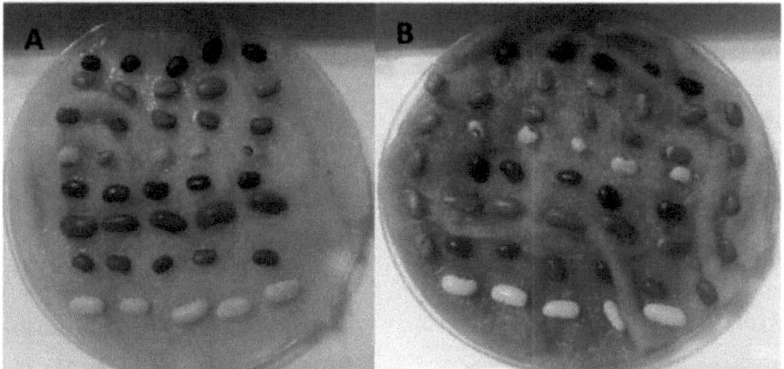

Figure 26. Germination test results for white, carioca, black-eyed, butter, brown, mulatto and red bean seeds in the presence of the flame retardant COUTOFLEX FIREKIL at a concentration of 9 per cent (A) and the flame retardant Phos Check G75R at a concentration of 13.4 per cent (B) after 14 days.

As observed in the simple test, the number of germinated seeds when using aqueous solutions of Coutoflex Firekill flame retardant is related to the concentration of retardant used in each solution. The graph below illustrates this relationship (figure 27).

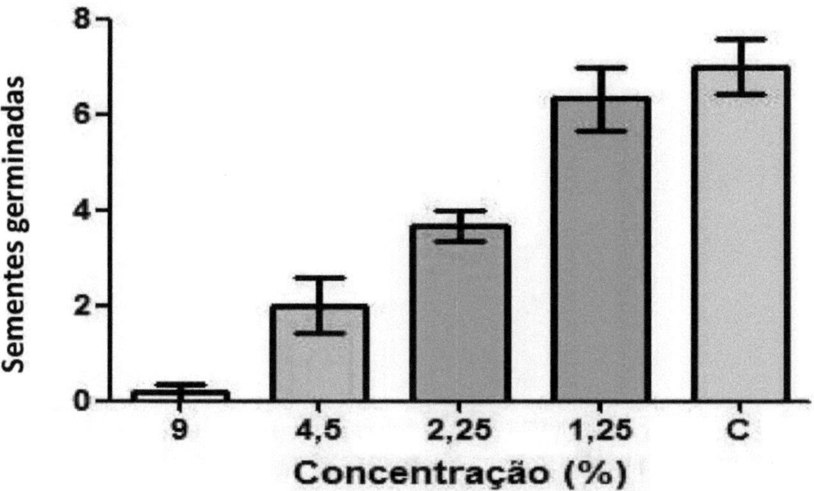

Figura 27. Relationship between the concentration of Coutoflex Firekill flame retardant and the number of germinated seeds. C: control without the product, germination rate between 91% ± 10%.

Based on the data obtained in the experiments carried out, both in simple conditions and in the laboratory, because the Coutoflex Firekill short-lasting flame retardant at a concentration of 1.125% (sample J) showed germination results very close to those obtained in the control samples tested, we decided to test its flame-fighting efficiency at this concentration. As the long-lasting flame retardant PHOS-CHEK G75R prevented the seeds from germinating at all the concentrations tested, we decided to discard the use of this retardant when optimising the Standard Operating Procedure for Combating Fire in Vegetation.

1.3. Experiment to evaluate the effectiveness of short-lived chemical flame retardants

The tests on lines 1, 2 and 3 were carried out to analyse the efficiency of the short-acting retardant at the 9% concentration indicated by the manufacturer, to serve as a parameter for comparison, as shown in Figure 28.

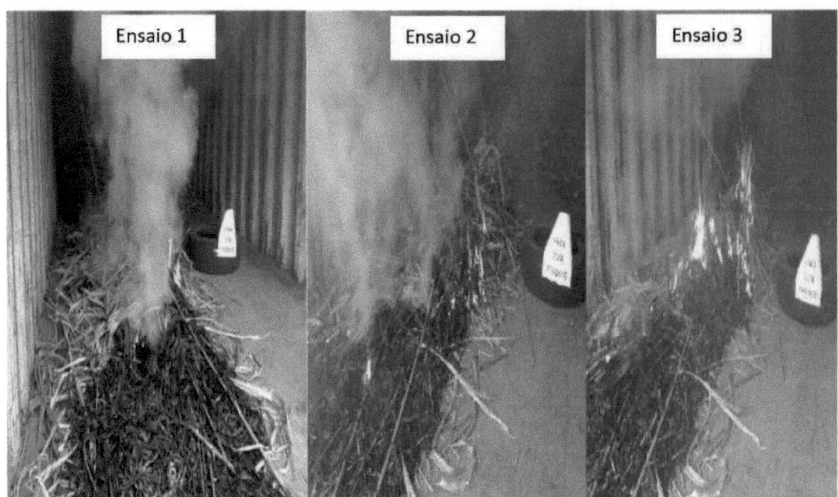

Figure 28. Tests 1, 2 and 3 of burning in an area without the application and an area with the application of Coutoflex Firekill flame retardant at 9% concentration.

Table 5 shows the fire propagation speed and flame length values in the areas with and without the application of the retardant product at a concentration of 9 per cent, collected during the burning of rows 1, 2 and 3, as well as the flame intensity values obtained using Equation 2, in addition to the mean values, variance and standard deviation of these parameters:

Table 5 - Fire propagation speed, flame length and flame intensity values obtained in area burning tests with and without Coutoflex Firekill flame retardant at 9% concentration.

	Speed of fire $(m.s)^{-1}$		Flame length (m)		Flame intensity $(kcal.m^{-1}.s^{-1})$	
	No product	With product	No product	With product	No product	With product
Test 1	0,0049	0,0015	0,52	0,32	15,36	5,35
Test 2	0,0051	0,0017	0,64	0,34	24,10	6,11
Test 3	0,0052	0,0018	0,61	0,31	21,72	5,00
Average	0,0050	0,0016	0,59	0,32	20,39	5,48
Variance	2,3E-08	2,3E-08	3,9E-03	2,3E-04	20,43	0,32
Standard Deviation	1,5E-04	1,5E-04	0,062	0,015	4,52	0,56

n=3

According to the values in Table 5, there was an average reduction of approximately

68 per cent in the speed of fire spread between the area with the product and the area without the product, while the flame length showed an average reduction of approximately 45 per cent between the two areas.

This reduction in flame length represented an average reduction of approximately 73 per cent in flame intensity, thus demonstrating the effect of the product applied.

Following trials 1, 2 and 3, trials 4, 5 and 6 were carried out (Figure 29). In these tests, the short-acting retardant was used at a concentration of 1.125% in order to check whether the product remained effective at this concentration.

Figure 29. Tests 4, 5 and 6 of burning in an area without the application and an area with the application of Coutoflex Firekill flame retardant at a concentration of 1.125%.

Table 6 shows the fire propagation speed and flame length values in the areas with and without the application of the retardant product at a concentration of 1.125%, collected during the burning of rows 4, 5 and 6, as well as the flame intensity values obtained using Equation 2, in addition to the mean values, variance and standard deviation of these parameters:

Table 6. Fire spread rate, flame length and flame intensity values obtained in area burning tests with and without Coutoflex Firekill Flame Retardant at 1.125% Concentration.

	Speed of fire $(m.s)^{-1}$		Flame length (m)		Flame intensity $(kcal.m^{-1}.s)^{-1}$	
	No product	With product	No product	With product	No product	With product
Test 4	0,0053	0,0027	0,56	0,43	18,04	10,17
Test 5	0,0047	0,0029	0,61	0,47	21,72	12,33
Test 6	0,005	0,0031	0,59	0,48	20,20	12,91
Average	0,005	0,0029	0,58	0,46	19,99	11,80
Variance	9E-08	4E-08	6,3E-04	0,0007	3,41	2,09
Standard Deviation	0,0003	0,0002	0,025	0,026	1,84	1,44

n=3

Analysing the values obtained in burning tests 4, 5 and 6 (table 6), there was a reduction of approximately 42% in the average speed of fire spread when compared to the value obtained for the area with the product applied.

With regard to flame length, there was a reduction of approximately 20 per cent and this resulted in a reduction of approximately 40 per cent in the average flame intensity.

The graphs below (figure 30, 31 and 32) illustrate the difference between the reductions in the average values of fire speed $(m.s^{-1})$, flame length (m) and fire intensity $(kcal.m^{-1}.s^{-1})$ using the short-lived retardant at a concentration of 9% and 1.125%.

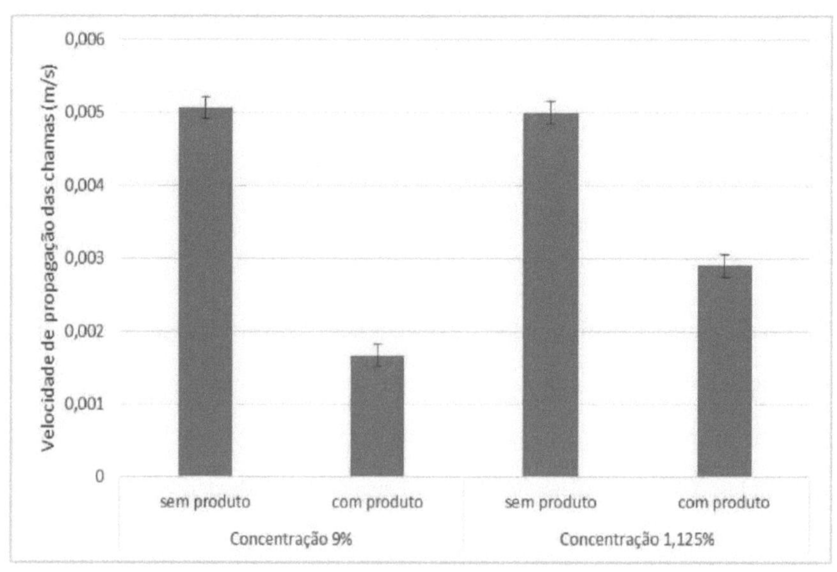

Figura 30. Graph illustrating the difference between the reduction in the average flame propagation speed values in the areas without application and with application of CoutoFlex Firekill short-lived flame retardant at 9% and 1.125% concentrations.

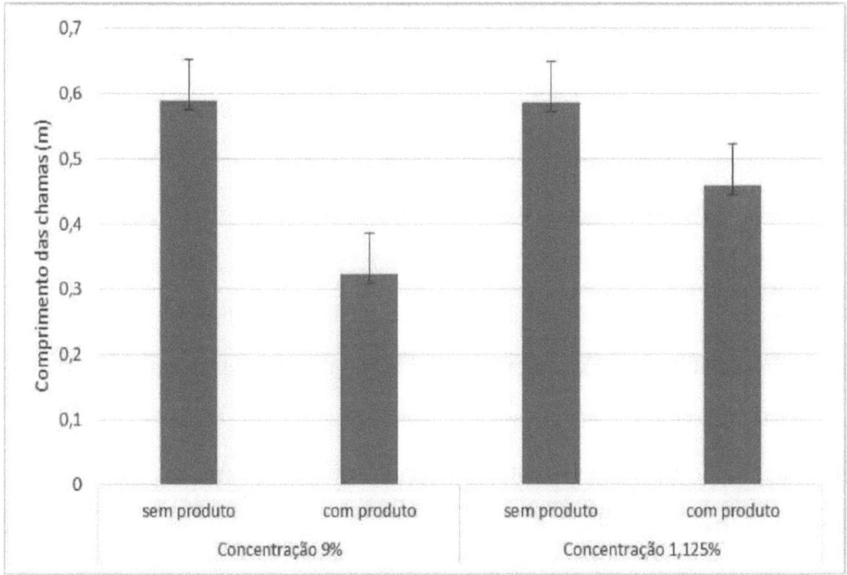

Figura 31. Graph illustrating the difference between the reduction in average flame length values in areas without and with the application of CoutoFlex Firekill short-lived flame retardant at 9% and 1.125% concentrations.

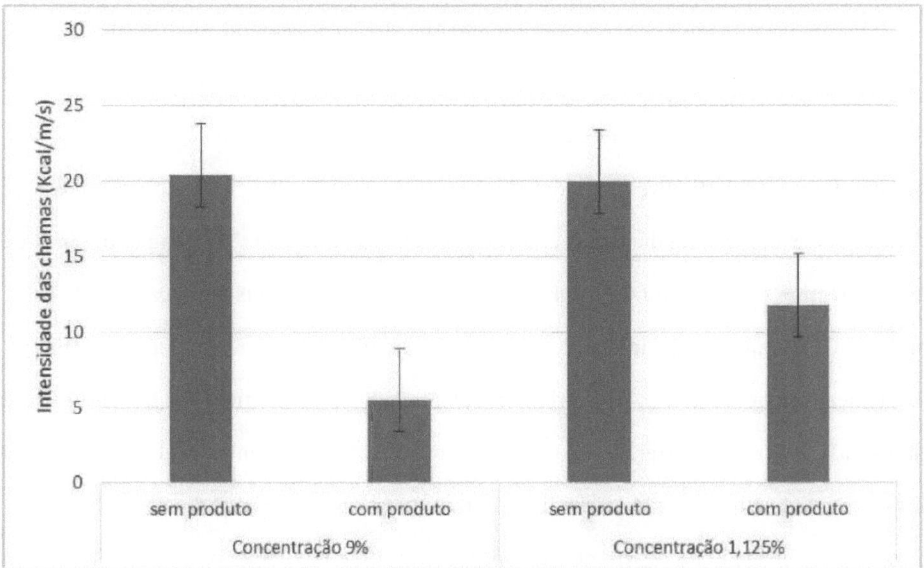

47

Figure 32. Graph illustrating the difference between the reduction in average flame intensity values in the areas without and with the application of CoutoFlex Firekill short-lasting flame retardant at 9% and 1.125% concentrations.

According to the data obtained in the burning tests (tables 5 and 6), it can be concluded that there was a decrease in the fire-fighting efficiency of the flame retardant when the concentration used was 1.125%. However, bearing in mind that fire in nature occurs first in undergrowth and its flames are normally no more than 2 metres long, the 20% reduction in flame length obtained for the 1.125% concentration would result in flames of approximately 1.5 metres, which would make it possible, according to PARIZOTTO (2006) and the Manual de un Primer Ataque a un Incendio Forestal (2002), to fight directly at the head or flanks using hand tools, thus making the product efficient even at this concentration. In addition, firefighters can use hand tools to cut and reduce the height of the vegetation before building the chemical firebreak, thus achieving flame lengths below 1.50 metres and, consequently, direct firefighting using hand tools.

5.4 Optimising the Standard Operating Procedure for fighting vegetation fires with a view to saving water resources

Analysing various procedures for fighting fire in vegetation, such as those of the Military Fire Brigade of the State of Rio de Janeiro (Annex I), the Military Fire Brigade of the State of Goiás, the Military Fire Brigade of the State of Mato Grosso do Sul and the Chico Mendes Institute for Biodiversity Conservation, in general, all the SOPs treat the phase of fighting and extinguishing the fire as a response phase with the aim of stopping or minimising the progression of the main front of the fire or the one that represents the greatest risk. None of the SOPs studied take into account the saving of water resources, so, if available, firefighters can arrive at the scene with firefighting vehicles and **carry out the fight using thousands of litres of water.** Just to give you an idea of the amount of water used to fight fires, according to the statistics department of the Operational Section of the Seventh Military Fire Brigade of the State of Rio de Janeiro, located in the municipality of Barra Mansa, in

2014 there were 236 vegetation fires that consumed **approximately 1.18 million litres of water** (figure 33).

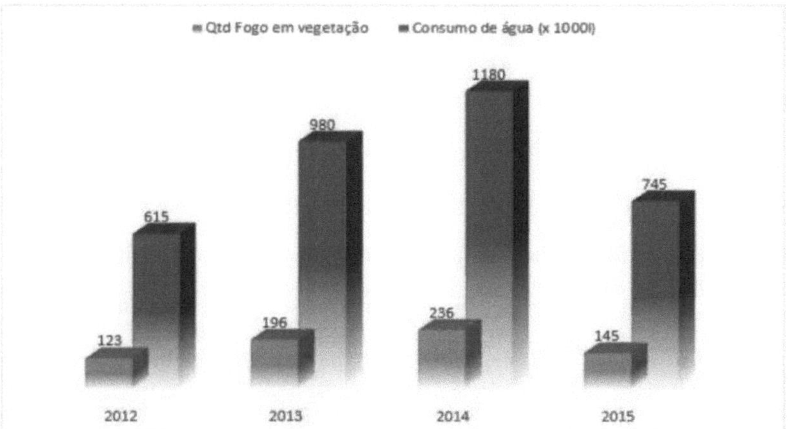

Figure 33. Graph illustrating the number of vegetation fires in the operational area of the Seventh Military Fire Brigade - Barra Mansa and the amount of water used to fight these fires.

Source: Operational Section of the 79GBM - Barra Mansa

As the aim of this work is to minimise the consumption of water used in vegetation firefighting. We will act mainly by changing this stage of the SOP.

5.4.1 Fighting and extinguishing vegetation fires while minimising the consumption of water resources

This is the phase in which the combatants begin the actions necessary to extinguish the fire. The firefighters must outline the objectives of the firefight, defining priorities, whether it's containing the spread, defending lives or defending property. The firefighting garrison must include at least 5 firefighters equipped with the necessary PPE, such as helmets, balaclavas, gloves and goggles, as well as equipment for fighting the flames, such as 02 back packs, 02 dampers and 01 sickle.

Before starting the operation, the spray bags should be filled with a mixture of water and CoutoFlex Firekill short-lived chemical flame retardant at a concentration of 1.125%.

Once the objectives have been outlined, the combatants must define the combat method to be used to resolve the situation.

Whenever possible, combatants should carry out direct combat with the dampers and bag tools, minimising the use of water resources.

In the case of fires with flame lengths of more than 1.5 metres, firefighters should use parallel firefighting, building a chemical firebreak, also called a cold line, at a safe distance from the perimeter of the fire, using backpack pumps containing a mixture of water and the short-lived chemical flame retardant, as well as scythe tools. The scythe should be used to cut down the vegetation before applying the aqueous solution containing the flame retardant. When the fire reaches this firebreak, its flame length will be reduced, and consequently so will its intensity. The firefighters will then be able to extinguish the fire using the dampers.

The members of the garrisons must take turns with the materials during the operation, in order to minimise wear and tear on the combatants.

During the experiment, 600 ml of the chemical retardant solution and water were used in each line to fight the fire in an area of $1.5m^2$. Thus, a 5-litre backpack is enough to make a chemical firebreak 1.5 m wide and 12.5 m long.

In the range where the product was applied, in all the burns carried out, combustion was slow, with small flames and a lot of smoke, a characteristic sign of the action of the retardant, which acts on the combustion phases, inhibiting the combustion of gases and intensifying the combustion of solids.

CHAPTER 6

DISCUSSIONS

6.1 Toxicity testing of COUTOFLEX FIREKILL AFFF and ARC **flame retardants** and PHOS-CHEK G75R **carried out simply in forest firefighting units and laboratories**

Analysing the results obtained in both the simple tests and the laboratory tests, it became clear that the long-lasting flame retardant Phos-Check G75 R has an inhibitory effect on the fermentation of the seeds of the test organisms used. For this reason, its use for optimising the SOP was discarded. It is necessary to carry out more in-depth tests to find out why the seeds didn't germinate even when in contact with aqueous solutions with very low concentrations of this flame retardant, in order to find out what structural changes the product caused in the test organisms. The short-lasting flame retardant Coutoflex Firekill, on the other hand, showed germination results very close to those of the control samples used when its concentration was reduced to 1.125%.

6.2 Experiment to evaluate the effectiveness of short-lived chemical flame retardants

During the experiment, average reductions of 20% in flame length were observed when we used the short-lasting flame retardant CoutoFlex Firekill at a concentration of 1.125%. We therefore decided to use this flame retardant at this concentration in order to optimise the Standard Operating Procedure (SOP) for fighting fires in vegetation, with the aim not only of extinguishing the fire, but also of significantly reducing the amount of water used to fight such fires.

Taking the Seventh Military Fire Brigade - Barra Mansa as a base (figure 33), assuming that in 2015 it were possible to adopt the method proposed in this dissertation to combat just 30% of the vegetation fires that occurred, it would be possible to save approximately 223,000 litres of water. Bearing in mind that this saving would only apply to one firefighting unit, if the method were adopted nationwide, millions of litres of water could

be saved.

52

CHAPTER 7

CONCLUSIONS

1.The long-lasting retardant PHOS-CHEK G75R, widely used in Brazil and freely sold on websites and in shops specialising in agricultural products, was shown to inhibit the germination of bean seeds (Phaseolus vulgaris) at all the concentrations tested. As such, its use in drawing up a Standard Operating Procedure (SOP) for fighting forest fires was discarded.

2.The short-lived retardant COUTOFLEX FIREKILL at a concentration of 1.125% showed germination results very close to those of the control samples tested. The CoutoFlex Firekill retardant showed satisfactory performance in firefighting, being able to reduce the flame length by 20 per cent, 42 per cent in the speed of fire propagation and promoted a 40 per cent reduction in flame intensity, even when used at a low concentration.

3.The SOP for fighting forest fires was optimised by using parallel firefighting, using an aqueous solution containing the short-lived flame retardant CoutoFlex Firekill at a concentration of 1.125%, with the aim of reducing the length of the flames to below 1.5 m, making it possible to use hand tools directly and thus reducing the amount of water used to fight the fire.

REFERENCES

NATIONAL CIVIL PROTECTION AUTHORITY. Operational Manual: use of air resources in civil protection operations. Lisbon: ANP, 2009. Available at: http://www.cm-evora.pt/NR/rdonlyres/0A2952A2-268C-4E58-A48CD985E965B42F/38487/ManualOperacionalempregodeMeiosAereos emopera%C3%A7%C3%B5esde.pdf. Accessed on: 15 February 2016.

ALEXANDER, M.E. (2000). Fire behaviour as a factor in forest and rural fire suppression. Forest Research, Rotorua, in association with the New Zealand Fire Service Commission and National Rural Fire Authority, Wellington. Forest Research Bulletin No. 197, Forest and Rural Fire Scientific and Technical Series, Report No. 5. 28 pp.

BATISTA, A.C. Detection of Forest Fires by Satellites. Revista Floresta, Curitiba-PR, V. 34, n.2, p. 237-241, May/August, 2004.

BATISTA, A.C. The use of retardants in aerial forest firefighting. Revista Floresta, Paraná, v. 39, 2009. Available at: http://www.floresta.ufpr.br/firelab/artigos/artigo424.pdf. Accessed on: 20 August 2015.

BATISTA, A.C. Models for estimating fire behaviour. Revista Floresta, Paraná, v. 37, 2007. Available at: http://www.floresta.ufpr.br/ firelab/artigos/artigo361.pdf. Accessed on: 09 August 2016.

BATISTA, A.C. et al. Evaluation of the efficiency of a long-lasting retardant, based on ammonium polyphosphate, in controlled burnings under laboratory conditions. Revista Sci. For, Piracicaba, v. 36, n. 79, p. 223-229, September 2008.

BOTELHO, H.S., et al. Educational meeting on forest fires. UTAD, Vila

Real, Portugal, 1996.

BYRAM, G.M. (1959). Combustion of forest fuels. In David K. P. (Ed.). Forest fire control and use (pp. 61-89). New York: McGraw-Hill Boock Company.

CASTRO, C.F. Combate a Incêndios Florestais, Portugal, V.13., p. 9, 2003.

CASTRO, C.F. et al. Forest Firefighting: initial training manual for firefighters. 3. ed. revised and updated. Lisbon: Sintra, 2006. 94 p.

CHANDLER, C.; CHENEY, P.; THOMAS, P.; TRABAUD, L.; WILLIAMS, D. Fire in forestry. New York: John Wiley, 1983. v.2, 298 p. (Forest Fire Management and Organisation).

CONSEJERÍA DE MEDIO AMBIENTE. Manual del Primer Ataque a un Incendio Forestal. Junta de Castilla y León, 2002

COUTOFLEX HOSE INDUSTRY. FIREKILL AFFF and ARC Foam Generating Liquid. Technical product sheet. Macaé, 2010. Unpublished work.

CRUZ, M.G., VIEGAS, D.X. (2001). Characterisation of fire behaviour in common fuel complexes in the Central Region of Portugal. Silva Lusitana, 9 (1), 13-34.

CUNHA, F.F. Chemical retardants in the fight against forest fires. Jovem Sul News, Chapadão do Sul, 27 October 2010. Available at:

http://www.jovemsulnews.com.br/user3/index.php?origem=colunista&xxx=1&id =1209&columnist=Fernando%20F.%20da%20Cunha. Accessed on: 24 August 2015.

CALADO, S.C.S. The water crisis and the availability of water for human needs. Journal of industrial chemistry, 1st quarter 2015. Available at: http://www.abq.org.br/rqi/2014/746/RQI-746-pagina4-Capa-A- crise-hidrica-e-a-disponibilidade-de-agua-para-as-necessidades-humanas.pdf.
Accessed on: 25 August 2015.

CBMERJ - MILITARY FIRE BRIGADE OF THE STATE OF RIO DE JANEIRO JANEIRO, Manual Básico de Bombeiro Militar, Rio de Janeiro, V.3, 2014.

CBPMESP - SÃO PAULO STATE MILITARY POLICE FIREFIGHTERS' BODY. Forest firefighting manual. São Paulo: PMESP,

2006. (Collection of Firefighters' Technical Manuals). Available at: http://pt.scribd.com/doc/55571856/Combate-a-incendios-florestais-SP.

DUFFUS, J.H. Glossary for chemists of terms used in toxicology (IUPAC recommendations, 1993). Chemistry Division Commission on Toxicily, v. 65, n.9, p.2003-2122, 1993.

FISKESJO, G. The Allium test as a Standard in environmental monitoring. Hereditas, v. 102, n. 1, p. 99-112, 1985.

GUARANY. PHOS CHEK® G75R: flame suppressant. Technical product sheet. São Paulo, 2011, 2 p. Unpublished work.

ICMBio - Chico Mendes Institute for Biodiversity Conservation. Ministry of the Environment. Available at: http://www.icmbio.gov.br/portal/ultimas- noticias/20-geral/8353-aceiros-ajudam-a-evitar-incendios-florestais. Accessed on: 16 August 2017.

INPE - National Institute for Space Research. Available at http://www.inpe.br. Accessed on: 05 January 2016.

INSTITUTO ALTO MONTANHA DA SERRA FINA, 2013. Available at:

http://institutoaltomontana.blogspot.com.br/2013/05/. Accessed: 17 August 2017.

MARTINS, S.D.R. Forest fires: Behaviour, safety and extinction. 2010. Dissertation (Master's Degree in Social Dynamics, Natural and Technological Risks) - University of Coimbra. Coimbra, Portugal, 2010.

MONSANTO COMPANY. Phos-chek: fire retardant in prescribed burning - Application guide. Publication number 9185 Supersedes IC/FP - 202 A.

MOTTA, D.S. Identification of the factors that influence fire behaviour in forest fires. 2008. 32 fls. Course Conclusion Paper (Higher Degree in Forestry Engineering) - Federal Rural University of Rio de Janeiro. Seropédica, 2008.

NUNES, J.R.S. FMA - A new forest fire danger index for the state of Paraná - Brazil. Curitiba: UFPR, 2005.150 p. Thesis (Doctorate in Forestry Sciences) - Federal University of Paraná.

OLIVEIRA, M. Manual of strategies, tactics and techniques for structural firefighting: Command and control of fire operations. Florianópolis: Editograf, 2005.

PARDO, J.M. Retardantes terrestres, uma novidad en La lucha de incêndios forestales. IV South American Symposium on preventing and fighting forest fires and 8ª Joint technical meeting SIF/FUPEF/IPEF on forest fire control. Belo Horizonte, 2007.

PARIZOTTO, W. Standardisation of procedures in forest fire control operations in the state of Santa Catarina. 2006. 62 fls. Monograph (Specialisation in Firefighting for Officers) - Military Firefighter Education Centre, Military Firefighter, Florianópolis, 2006.

PIAUÍ. State Secretariat for Rural Development. São Francisco and Parnaíba Valley Development Company. Course booklet on techniques for preventing and fighting forest fires. Curitiba: STCP, 2010. Available at: http://www.codevasf.gov.br/search?SearchableText=apostila.

PMESP. POLÍCIA MILITAR DO ESTADO DE SÃO PAULO - CORPO DE BOMBEIROS, Terminologia de segurança contra incêndio, 2011.

PMESP. POLÍCIA MILITAR DO ESTADO DE SÃO PAULO - CORPO DE BOMBEIROS, Manual de Fundamentos do Corpo de Bombeiros, 2nd edition, 2006.

RIBEIRO, G.A. Air combat technologies and the use of retardants. Opiniões magazine, March/May 2011. Available at: http://ojs.c3sl.ufpr.br/ojs2/ index.php/floresta/article/view/13160/8907.

UN - UNITED NATIONS FOOD AND AGRICULTURE ORGANISATION. Terms and definitions. Italy 2015 p.3. Available at: http://www.fao.org/docrep/017/ap862e/ap862e00.pdf.

PARIZOTTO, W. Standardisation of procedures in forest fire control operations in the state of Santa Catarina. 2006. 62 fls. Monograph (Specialisation in Firefighting for Officers) - Military Firefighter Education Centre, Military Firefighter, Florianópolis, 2006.

PEREIRA, J.F. Variation in the humidity of forest fuels as a function of the fire danger indices fma and fma+ in a Pinus elliottii stand in the municipality of Rio Negro - PR. 2009. Dissertation (Master's Degree in Forestry Sciences) - Federal University of Paraná, Curitiba, 2009.

RIBEIRO, G. A. et al. Efficiency of a long-lasting retardant in reducing the spread of fire. Revista Árvore, Viçosa-MG, v. 30, n. 6, p. 1025-1031, dec. 2006. Available at: http://www.scielo.br/scielo.php?script=sci_ arttext&pid=S0100676220060006000018&lng=en&nrm=iso&tlng=en

RODRIGUES, A.N.C. Considerações sobre prevenção e combate aos incêndios florestais no Estado do Rio de Janeiro. 2008. 32 fls. Course Conclusion Paper (Higher Degree in Forestry Engineering) - Federal Rural University of Rio de Janeiro. Seropédica, 2008. Available at: http://www.if.ufrrj.br/inst/monografia/2007II/Aline%20Nahanna%20Carneiro %20Rodrigues.pdf.

SCHUMACHER, M.V.; BRUN, E.J.; CALIL, F.N. CADERNO DIDÁTICO: CFL 506 - PROTEÇÃO FLORESTAL. Santa Maria: Federal University of Santa Maria, 2005. Available at: http://w3.ufsm.br/labeflo/ensino/graduacao/ protecao/caderno_2005.pdf.

SANTA CATARINA. Decree no. 4909 of 9 October 1994. Establishes fire safety regulations for the state of Santa Catarina.

SANTANNA, C.M.; FIEDLER, N.C.; MINETTE, L.J. Forest **fire control.** Alegre, ES. Os Editores, 2007. 152 p.

SANTOS, N.R. Notions of surface forest firefighting. 2009. 92f. Course Conclusion Paper

(Technology in Emergency Management) - Land and Sea Technology Centre, Vale do Itajaí University, São José, 2009.

SAVIOLI, L.H.. Fighting forest fires with aircraft. 1998. 162 f. Thesis (Master's Degree - Officer Training) - Centre for Training and Higher Studies - PMSP, São Paulo, 1998. Available at: http://pt.scribd.com/doc/555008856/Combate-aincendios-florestais-SP.

SILVA, R.G. Manual for Preventing and Fighting Forest Fires. Brasília, 1998.

SOARES, R.V.; BATISTA, A.C. Distance learning specialisation course: biomass combustion and fire propagation. Brasília: UFPR - Brazilian Association of Higher Agricultural Education, 35 p. 2 v. 2006.

SOARES, R.V.; BATISTA, A.C.; SANTOS, J.F. Evolução do perfil dos incêndios florestais em áreas protegidas no Brasil, de 1993 a 2002. Blumenau, Proceedings of the II Seminar on Current Issues in Forest Protection. P. 10, 2005.

SOARES, R.V.; BATISTA, A.C. Incêndios Florestais: Controle, efeitos e uso do Fogo. Curitiba: FUPEF, 2007. 298 p.

SOARES, R.V.; SANTOS, J.F. Profile of forest fires in Brazil from 1994 to 1997. Revista Floresta, Curitiba, V.32, n.2, p. 219-232, 2002. Available at: http://ojs.c3sl.ufpr.br/ojs2/index.php/floresta/article/view/2287

SOARES, R.V. Profile of forest fires in brazil in 1983. Brasil Florestal no. 58: 31-42. 1984.

SOLO. Original chainsaw manual. Sindelfingen, 2008. Available at: http://www.sologermany.com/gba_download/9646100/web/9646100_pt_web_0 8_2008.pdf.

STANGERLIN, D.M. et al. Quantification of combustible material accumulated in the litter of a Eucalyptus grandis forest. In: CONGRESSO DE INICIAÇÃO CIENTÍFICA, Santa Maria: Faculdade de Agronomia Eliseu Maciel, 2007.

SUASSUNA, F.F. Standard Operating Procedure - Fire in Vegetation. Military Fire Brigade of the State of Rio de Janeiro, 2013.

Tomé M. and Borrego C. (2002) Fighting wildfires with retardants applied with airplanes. Forest Fire Research & Wildland Fire Safety, Viegas (ed.), Coimbra.

VÉLEZ, R.M. Defence against forest fires. Madrid: McGraw Hill, 2000.

VIEGAS, D.X. Fire behaviour and personal safety. Proceedings of Jornada de Prevencion de riegos laborales y ambientales, Seville, Spain. 2006.

I want morebooks!

Buy your books fast and straightforward online - at one of world's fastest growing online book stores! Environmentally sound due to Print-on-Demand technologies.

Buy your books online at
www.morebooks.shop

Kaufen Sie Ihre Bücher schnell und unkompliziert online – auf einer der am schnellsten wachsenden Buchhandelsplattformen weltweit! Dank Print-On-Demand umwelt- und ressourcenschonend produziert.

Bücher schneller online kaufen
www.morebooks.shop

Printed by Books on Demand GmbH, Norderstedt / Germany